The crayfish:
its nature and nurture

Cover: The common crayfish (*Astacus fluviatilis*). The drawing used as a frontispiece of 'The Crayfish. An introduction to the study of zoology' T. H. Huxley, F.R.S. in 1880.

The crayfish:
its nature and nurture

Roy E. Groves

Fishing News Books Ltd
Farnham · Surrey · England

© Roy E Groves 1985

British Library CIP Data

Groves, Roy E.
    The crayfish: its nature and nurture
    1. Crayfish   2. Aquaculture
    I. Title
    639'.541   SH3809

ISBN 0 85238 137 9

Published by
Fishing News Books Ltd
1 Long Garden Walk
Farnham, Surrey, England

Typeset by
Alresford Phototypesetting
Alresford, Hants

Printed in Great Britain by
Whitstable Litho Ltd
Whitstable, Kent

Contents

Figures

Abstract

There are plenty of sources of scientific information provided by learned bodies, biologists and zoologists on the subject of the freshwater crayfish, its biology, structure, life style and habitat. There is a lack of published material written in everyday language for the ever increasing number of people interested in this fascinating creature and its culture.

This book tries to fill that gap. The first part describes the animal, its main characteristics, reproductive processes, growth, mode of life and its life in the wild. The second part poses the problems to be encountered in the intensive culture of the crayfish, as yet largely a matter of speculation, and suggests some ways by which these problems can be overcome.

Since the crayfish is almost unknown in the United Kingdom as a food, although increasing foreign travel is making it more familiar, there is a chapter devoted to simple cooking and smoking methods.

Acknowledgements

I am most grateful to the many people who introduced me to the crayfish and who contributed to my knowledge of this creature including Alex Behrendt of Two Lakes, Hampshire, Dr David Holditch of the Department of Zoology, The University of Nottingham, Dr K Bowler of the Department of Zoology, The University of Durham, John Hogger of Thames Water, Reading, John Wickins of the Fisheries Experiment Station, Conway, Michael Brown of Brown and Forest Crayfish Supplies, Drayton, Somerset, Mark Mount of Littlebourne, Kent, Mr & Mrs R White of Fyfield Hall, Ongar, Essex and many others who allowed me to see their activities in this field. I also acknowledge the valued contribution made to current thinking on this subject by Dr J S Goddard and the authorities of the Hampshire College of Agriculture through the periodical crayfish courses and the production of the Crayfish Bulletin.

Roy E Groves

Part one
THE ANIMAL

Introduction

The crayfish is an animal of the Class *Crustacea*, Sub-class *Malacostraca* and Order *Decapoda*. It is an animal with four pairs of walking legs as opposed to the swimming decapods such as shrimps. There are many species found naturally in many different parts of the world from the very large Marron (*Cherax tenuimanus*), the giant crayfish of Australia, to the comparatively small animal which is found naturally in the rivers and streams of England (*Austropotamobius pallipes*).

Huxley (1881) said that the origin of the word crayfish involves curious questions of etymology. The old English method of writing it was 'crevis' or 'crevice', cray being a phonetic spelling of 'cre' whilst 'fish' is the 'vis' insensibly modified to suit our knowledge of the creature as an aquatic animal. 'Crevis' is one of two things, either a modification of the French 'écrevisse' or the low Dutch 'crevik'. He considered that the former is the more usually accepted one and therefore concludes that, since it joins the other French equivalents such as mutton for sheep's flesh and beef for ox flesh, we would not have called a crayfish a crayfish had it not been for the Norman Conquest. If however 'crevik' is the correct source of the word it may have come straight from the Anglo-Saxon part of our mixed ancestry.

The crayfish has always been recognized as a source of food and for years has been trapped, caught with drop nets, fished for by line or simply hunted in streams and caught by hand. It is only in recent years that serious attempts have been made in Europe to rear the crayfish commercially for the table. In France, for example, the crayfish is considered a great delicacy, imports reaching some 2,000 tonnes each year as compared with the five or so tonnes brought into the United Kingdom. The principal exporting country is Turkey which annually ships some 4,000 tonnes of native crayfish. Turks, being Moslems, are precluded for religious reasons from eating the crayfish. So large are the catches from the wild that it is reported that over-fishing is taking place with possibly disastrous results. Greece, Italy, Yugoslavia, Kenya, Uganda, the USA and more recently Spain are other exporting countries but the amounts are very small. Comparative data for annual consumption, production, imports and exports are shown in Table 1. There is an extensive industry in Louisiana, USA, where vast expanses of marshland are alternately drained and flooded to produce considerable numbers of crayfish.

The crayfish was well known in schools, being one of the creatures used extensively for dissection by students of biology. This practice was abandoned some years ago.

Most attempts to raise crayfish commercially, that is to say in the way that trout are farmed, have foundered because of the inherent

Table 1. COMPARATIVE DATA FOR ANNUAL CRAYFISH PRODUCTION AND CONSUMPTION IN EUROPE.
(*Based on a table produced by J. V. Huner*)

Country	Total consumed	Imports	Domestic production	Exports
Sweden	2100+	2000+	100+	0
France	2010+	2000+	10+	0
West Germany	?	200	?	–
Turkey	0	0	4000+	4000+
Spain	1800+	0	2000+	200+
Finland	110+	50 – 60	60 – 120	3
Norway	10 – 15	0	20 – 30	10 – 15
Denmark	40+	40+	?	?
Belgium	150+	150+	?	?
Switzerland	30 – 40	30 – 40	?	?
United Kingdom	7	5	2	0
Totals	6272	4495	6260	4228

Note This table as amended can only be considered as approximate. The figures from West Germany are highly suspect, both Professor Huner and the author got very little information from the authorities there. Belgium and France serve as import ports for other Western European countries and percentages shipped elsewhere are not known.

difficulties which are encountered and which have not yet been overcome. These stem from the nature of the animal, itself cannibalistic, such as its susceptibility to disease, and especially to the large and varied numbers of predators. It has been estimated that, in the wild, less than 10% of those hatched reach maturity.

Of all the species of crayfish there can be no doubt that the Signal crayfish (*Pacifastacus leniusculus*), so-called because of the very distinctive colours – red, blue and white on its great claws which are particularly noticeable when it stands in its aggressive position – has the most potential for farming in temperate climates. This crayfish, which originated in California, has been extensively introduced into Europe since 1960 when it was imported into Sweden to replace the native crayfish population which had been devastated by the plague (*Aphanomyces astaca*) and by the so-called Porcelain disease (*Thelohaniosis*) which are described in a later chapter. It is now found in Finland, the USSR, Germany, Austria, Luxembourg and France. It was originally brought to England from Sweden but there are now sufficient first generation stocks in this country to render further imports unnecessary. It is to be remembered that under the Wild Life and Countryside Act 1983 it is illegal to release into British waters any crayfish other than the native species (*Austropotamobius pallipes*). The Signal crayfish is purported to be plague resistant, but it may well be that it can act as a carrier and there was much opposition to its introduction by conservationists and others fearing that it would oust the native creatures. There are so many kept today in open ponds, reservoirs *etc* that any damage has almost certainly been done since individuals will move far from their original site especially if kept in overcrowded conditions. They can travel for many miles not only in water but across land.

The principal advantage of the Signal crayfish to the farmer is in its greater size and its more rapid growth under ideal conditions. Its claws are larger than any other temperate zone crayfish. To illustrate size, *Figs 1* and *1a* show Signal crayfish and other crayfish of the same age.

Other species found in Europe are:

The American River Crayfish (*Orconectes limosus*) found mainly in Germany and Eastern Europe.

The Red Swamp Crayfish (*Procambarus clarkii*) the crayfish

Fig 1 A signal crayfish (Pacifastacus leniusculus) and a native crayfish (Austropotamobius pallipes) of the same age. *John Hogger*

extensively raised in the swamplands of Louisiana. It has been introduced recently into the rice paddy fields in Spain. It is also found in Kenya, Uganda, Japan into which it has been imported and experimentally in France and England.

The Crayfish of the Streams (*Austropotamobius torrentium*). It is doubtful whether there are any still to be found.

The White-footed Crayfish (*Austropotamobius pallipes*). This is the only native crayfish found in the United Kingdom, and, in its several varieties, in many parts of Europe, especially in France.

The Slender-clawed or Turkish Crayfish (*Astacus leptodactylus*). Found extensively in Turkey and the great river basins of Eastern Europe – the Volga, Don and Danube. It has been reported that one of these was recently found alive in a lake in the south of England – presumably a live specimen from Billingsgate fish market introduced by its purchaser.

Fig 1a A signal crayfish (Pacifastucus leniusculus) and a noble crayfish (Astacus astacus). *Top* of the same age and *bottom* of the same body length. *J. Cukerzis*

13

The Red-footed or Noble Crayfish (*Astacus astacus*). The species which was obliterated in Sweden and replaced with the Signal crayfish. Found in many parts of Europe from the Baltic to Northern Italy. Being very susceptible to disease and the effects of pollution, it is gradually disappearing although there have been attempts recently to introduce this species for culture purposes.

There are nearly one hundred species of crayfish to be found in Australia, two of which, the Marron (*Cherax tenuimanus*) and the Yabby (*Cherax destructor*) are farmed, mostly in irrigation dams and ponds.

General description

The crayfish is a small, freshwater, lobster-like creature which in nature inhabits ponds, streams and rivers. It has a body encased in a hard shell composed mainly of a substance called chitin which is an acetyl-glycosamine ($C_{32}H_{54}O_{21}N_4$). This is called the cuticle or exoskeleton and is shed at the time of moult.

The body is divided into three main parts (*Fig 2*), the head, the thorax and the abdomen. The head and thorax are almost completely fused together to form what is called the cephalothorax. Where they are joined there is a crease or groove which is known as the cervical groove.

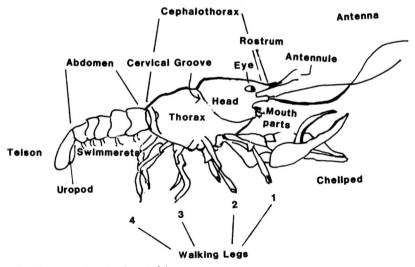

Fig 2   The principal parts of a crayfish

15

It is a segmented creature divided into twenty segments, each carrying a pair of jointed limbs or appendages. Each of these has a vital role to play in the life of the animal – vision, touch, smell, taste, feeding, breathing, locomotion, defense and attack, mating, egg-laying and the nurture of the eggs and the young until free-swimming. A summary of the segments, their appendages and their functions is set out in Table 2. It will be noted that the first segment carries the eye stalks and these are not generally considered as appendages.

The appendages are jointed in two, three or four places and the main sections so formed are further sub-divided. This division and sub-division is clearly illustrated in *Figs 3* and *4* which show the make-up of each appendage and how these have similar forms of con-

Table 2. THE SEGMENTS AND THEIR RELATED APPENDAGES
WITH THEIR FUNCTIONS

Segment	Name	Functions
1	Preantennal. Two ocular peduncles	Vision
2	Antennule	Balance, touch, smell, taste
3	Antenna	Touch, smell, taste
4	Mandible	Feeding
5	1st Maxilla	Part of mouth
6	2nd Maxilla	Part of mouth
7	1st Maxilliped	Respiratory, feeding
8	2nd Maxilliped	Respiratory, feeding
9	3rd Maxilliped	Respiratory, feeding
10	Cheliped	Fighting, holding. The large claw
11	1st Walking leg (periopod)	Walking, holding. The small claw
12	2nd Walking leg (periopod)	Walking, holding. Small claw. Genital opening ♀
13	3rd Walking leg (periopod)	Walking. No claw
14	4th Walking leg (periopod)	Walking. Sexual opening ♂
15	Swimmeret (pleopod)	Circulation of water ♀ Seminal channel ♂
16	Swimmeret (pleopod)	Circulation of water ♀ Egg carrying ♀ Seminal channel ♂
17	Swimmeret (pleopod)	Circulation of water ♀
18	Swimmeret (pleopod)	Egg carrying ♀
19	Swimmeret (pleopod)	Egg carrying ♀
20	Paddle (uropod)	Locomotion. With telson forms tail fan. Protection of eggs ♀

16

struction but are modified or have parts omitted in accordance with their function. At the rear of the animal is the telson which is not a segment but which, together with the rearmost appendages, the uropoda, forms the fan-shaped tail.

At the front end of the head, there is a prolonged piece with a point which is called the rostrum on each side of which are found the eyes. The measurement of a crayfish is taken from the tip of the rostrum to the end of the tail.

The gill chambers of the creature are located in an area on the side

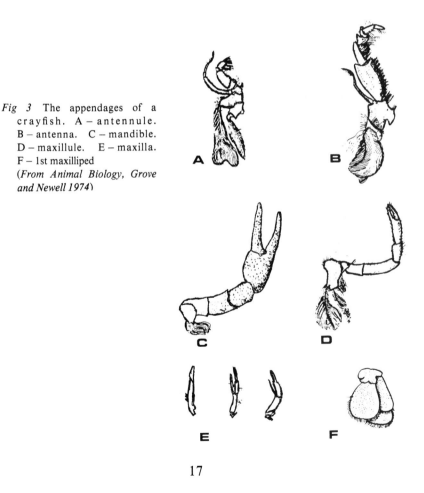

Fig 3 The appendages of a crayfish. A – antennule. B – antenna. C – mandible. D – maxillule. E – maxilla. F – 1st maxilliped
(*From Animal Biology, Grove and Newell 1974*)

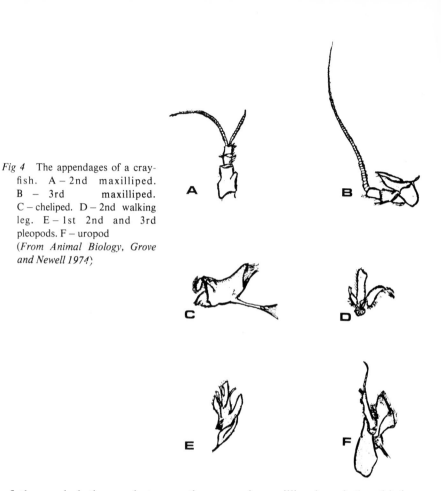

Fig 4 The appendages of a crayfish. A – 2nd maxilliped. B – 3rd maxilliped. C – cheliped. D – 2nd walking leg. E – 1st 2nd and 3rd pleopods. F – uropod
(*From Animal Biology, Grove and Newell 1974*)

of the cephalothorax between the second maxilliped and the third walking leg. The anus is situated on the underside of the telson.

These are the principal external parts of the animal. The internal organs and lay-out of the crayfish will be well-known to older students of biology since it was one of the main subjects for dissection. For this reason details of its muscular, alimentary, digestive, circulatory, cardiac, respiratory, excretory, nervous, sense and reproduction systems are set out and discussed at great length in older biology text books, some of which are referred to in the bibliography at the end of this book. They are of little interest to those responsible for the breeding and farming of crayfish except where they bear directly on external conditions necessary for their reproduction and growth.

Movement

Unlike its relatives the shrimps, the crayfish cannot swim. Its normal method of locomotion is by walking using its four pairs of walking legs for this purpose. When frightened the animal resorts to its escape mechanism which is a jerky darting movement at great speed *backwards*. It does this by means of its spread out fan tail and rear segments of the abdomen operated by six groups of muscles divided into extensors and flexors. These are attached in pairs to the fore end of the cephalothorax and the solid covers of the segments in the abdomen at the rear. The Signal crayfish in particular is much more agressive than most species and will sometimes stand instead of fleeing, facing the source of danger with its claws outspread in an attack position. The claws in this species are also more articulated allowing the animal to attack behind as well as in front of its body, which is the usual way in most crayfish. During its movements the crayfish keeps its antennae continually in action exploring the water all round it. When using its escape mechanism these are stretched along its body to detect any obstacles in the rear.

Feeding

The crayfish is omnivorous in that it will eat both animal and vegetable matter, both living and dead. Thus it is to some extent a scavenger, although, contrary to many beliefs, it does not deliberately go out of its way to find dead material. There is a tendency in younger specimens to eat more animal matter whereas the older adults will prefer a more vegetable diet. The crayfish is very circumspect when feeding and will examine, probe, taste and finally eat. Even when satisfied with the food it will return to its hide from time to time and then come back to finish off whatever it was eating.

Favourite live foods are various molluscs but they will, in the wild, eat caddis flies, larvae of all kinds, midges, worms, snails, tadpoles and even small rodents, birds and fish. They eat practically all aquatic plants with a special fondness for blanket weed. Their feeding habits are almost nocturnal. They will commence their search for food at about midday and continue until the early hours of the morning.

Habitat

The crayfish always lives in a 'hide'. This may consist of a hole in the bank, a crevice in the rocks, a shelter under the roots of trees or under artificial conditions in holes in brick rubble or other similar man-made caches, anywhere in fact where it feels able to defend itself against its many adversaries. Some species will burrow into the mud or into the sides of dams, banks of ponds *etc* sometimes causing great damage in the process. It will venture forth from this hide to find food, at times of moult and reproduction, but it knows that it is very vulnerable on all of these occasions and it will return to its hide as soon as it possibly can.

Various species of crayfish like different types of environment, some fast-flowing clear water, others deep and sluggish pools or swampland. There are, however, restrictions – the crayfish cannot survive in soft water, which is why there are no native crayfish in Wales. They need water which is reasonably well oxygenated. The optimum conditions are pH values of 7 upwards with 7 – 9 preferred; hardness of between 100 and 350 ppm $CaCO_3$ (crayfish need calcium since they moult and have to replace their exoskeletons); a minimum level of 6 ppm of dissolved oxygen. The annual water temperature range must include a minimum period of at least three months in the summer of 15 degrees Centigrade but there must be a drop for mating and for egg development. Reports indicate that extremes of cold in the winter do not adversely affect the crayfish found in Northern Europe. Ponds which are surrounded with trees and other vegetation which cause organic waste deposits at the bottom are generally unsuitable. It should be noted that if the environment is unsuitable or it is too over-crowded the crayfish will leave it, sometimes travelling great distances overland to find a new habitat. Although crayfish will leave the water to eat grass and other vegetation, if they are found some distance away from their habitat this probably means that it is unsuitable for them and it should be checked to ascertain the reason.

Reproduction

The sexing of crayfish

Once the animal has become sexually mature, in the case of the Signal crayfish at about two years of age, the identification of the sexes is comparatively easy and can usually be carried out with the naked eye although the use of a magnifying glass may be of assistance. In the female the oviduct openings are to be found at the base of the 2nd pair of walking legs (*Fig 5B*) whereas in the male (*Fig 5A*) the first and second pairs of swimmerets are adapted to provide the channels used for the transfer of the sperm to the female from the genital opening which can be found beneath them.

Mating

If, in the summer months, one bends the tail of a female crayfish one can clearly see just above the first abdominal segment and through the transparent membrane which protects them, the red-coloured mass of ovules. In the autumn from the end of September until November the period of copulation takes place. It is a slow process which often proves fatal to the female and during which either can be severely mutilated. The male hunts the female and ruthlessly attacks her. Sometimes she escapes. When he succeeds in catching her he will turn her over by means of his large strong pincers and they will interlock using their walking legs some of which have small claws or chela (*Fig 5B*). Several attempts may be necessary to achieve this during which time the risk of mutilation is high, but ultimately the male will master the female and will deposit the sperm-carrying liquid on the underside of the female on the plates surrounding the external orifices of the oviducts (*Fig 6*). This seminal fluid jellifies into strings

Fig 5 Sexing the crayfish
A – male showing the genital
opening – a.   B – female
showing the oviduct
opening – b

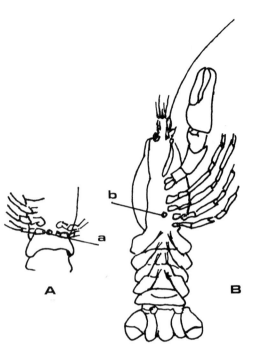

A   B

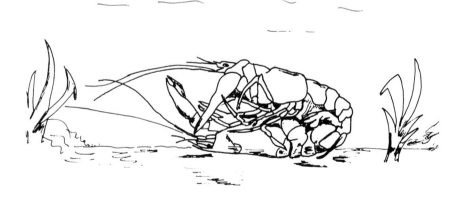

Fig 6   Copulation

22

of milky material; these threads are the sperm carrying tubes. There are two factors which have some bearing on the commercial farming of crayfish. Firstly it will be obvious that the chances of successful mating are higher if the partners are of similar size. A small male, no matter how ambitious, could not turn over a large female. Secondly one male can mate successfully with several if not many females. It is therefore suggested that in a breeding stock pond, one male to every three or four females is the right proportion.

Gestation

After copulation the female enters the periods of gestation. There are two such periods, the first when the eggs are retained within the mother and the second or 'open' gestation when she will carry the eggs externally. During the first period the female will remain in her hide or cache, leaving only for food, to allow the internal gestation to take place. This takes from four to six weeks. Towards the end of this period white blotches will appear along the underneath joints between the abdominal segments. At the actual emission of the eggs the female has to leave her hide to carry out various motions to assist with this and this is a period of great danger since it occurs when she has very little resistance and she is most vulnerable to attack. She folds herself into an arc with the tail spread to form a pocket which is then coated with a transparent fluid. The oviducts force out the eggs together with a fluid which dissolves the jellified male sperm carrying tubes. The sperms are liberated and the eggs fertilized. The transparent fluid which formed the pocket now becomes a membraneous substance which covers the eggs, isolating one from another, and attaching them firmly to the hairs which are to be found on the swimmerets and other parts of the abdomen of the mother (*Fig 7*).

The second period of gestation now begins. The female returns to her hide where she will circulate water over the eggs with the use of her swimmerets and tail and thus irrigate them. Unfertilized eggs will degenerate and fall off and she will clean, comb and rake the bunches of eggs to remove debris, sediment or impurities. This is a period of incubation and it varies according to the species and to the ambient temperature of the water. This can be assessed using what are known as *degree/days*. In the case of the Signal crayfish it has been established that it varies from 900 degree/days at 18°C to 1620

Fig 7 A female crayfish in berry. *John Hogger*

degree/days at 7°C. (That is to say a period of 50 days at the former temperature to 230 days in the colder environment.)

The eggs, which were blackish at the time of emission gradually take on a purplish colour and become transparent, the embryo being visible to the naked eye. It becomes active within its shell and rotates the cord with which it is attached to the mother. This cord carries out a mechanical function of attachment only and has no biological functions. All this time the mother will continue to keep the eggs clean and keep a constant flow of water over them. The number of eggs which will be carried in this manner will vary considerably with the quality of the environment and the availability of food. This number can be anything from 50 to 500 but with the Signal crayfish the average should be 200. This means that provided the survival rate can be maintained at a high level large numbers of breeding stock are not necessary.

The hatch

The hatch or eclosion usually takes place at the end of Spring. The mother will rake the eggs more vigorously. At the actual moment of the hatch, the egg will split and the young will appear. The early days of the crayfish, shown in *Table 3*, are unlike those of other decapods since the newly hatched creature bears a close resemblance to the adult crayfish whereas most others have distinct larval forms which require many changes to reach the adult stage. Immediately after the eclosion the tiny crayfish attaches itself to its mother by means of its claws which have their ends curved into small hooks for this purpose. It will remain thus attached for some time but will soon venture forth, rushing back when there is the slightest sign of danger. Together with all its brothers and sisters the crayfish will remain like this until its first moult when they will leave the mother completely. It is at this stage that they become very vulnerable to the cannibalistic instincts of the mother who will devour them if she gets a chance. Under controlled circumstances the mother must be removed from the proximity of her offspring as soon as they are free moving.

Table 3. CHARACTERISTIC FEATURES OF THE FIRST POSTEMBRYONIC STAGES OF *Astacus astacus*
(*After C. Payen, 1973*)

	Stage 1	Stage 2	Stage 3	Stage 4	Stage 5
Method of life	Joined to the mother by means of a thread (embryonic cuticle) attaching their tail to the empty egg-sac itself fixed to the mother's pleopods	Attached to the pleopods of the mother by the claws of the first walking legs and living separate at the end of Stage 2	Active, walks and swims	Ditto	Ditto
General bearing	Inactive larva, clustered Rostrum bent in Carapace spherical No hairs	Abdomen unfurled towards the back Hairs beginning to form on the antenna, walking legs and swimmerets	Little difference to that of a mature crayfish	Ditto	Ditto
Yolk	Already present to a greater or less extent	Disappears progressively	Absent	Ditto	Ditto
Feeding	None	Commences to feed at the end of this stage	Feeds	Ditto	Ditto
Position of the Antennae	Folded back along the body	Stretched out in front	Ditto	Ditto	Ditto
Eyes	Inset and immobile	On stalks (**Peduncles**)	Ditto	Ditto	Ditto

Pleopods 1	Small protuberances symmetrical on each	Ditto	Ditto	Ditto	Ditto on female. Differences in the stylets on male
2 3 4 5	Present and branched	Ditto	Ditto	Ditto	Branched in both sexes
Uropod	Absent	Ditto	Beginning to appear free and fan-shaped	Present	Ditto
Telson	Stretched out in a paddle shape and seamless	Folded back under the body	Stretched out	Lightly folded	Ditto
Genital openings Female	Not distinguishable	Not distinguishable	Probably present but hidden by a membrane	Not visible	Visible with difficulty
Male			Not visible	Ditto	Genital papilla not distinguishable

ω —— St.1 —— St.2 —— St.3 —— St.4 —— St.5 —— St.6 —— St.7 —— St.8

Hatch E1 E2 E3 E4 E5 E6 E7

Postembryonic moults

—— = 8 days

E1 — E7 = Moults

27

Growth

At the beginning of its life the tiny crayfish feeds on reserves which it carries in its cephalothorax which is why this appears out of proportion to the rest of the body. Its cuticle is pliable and soft which allows growth to take place. Soon however the cuticle hardens and forms what is known as the exoskeleton of a juvenile crayfish and in up to ten days after being hatched, the young crayfish leaves this exoskeleton through a split between the cephalothorax and the first abdominal segment. This is known as the moult or exuviation and henceforth the crayfish has to shed its exoskeleton at intervals and produce a new one. It is whilst the new one is soft that the body is able to grow rapidly. It follows therefore that moults are more frequent the younger the animal and this has been observed and recorded. The early postembryonic days of the crayfish are set out in *Table 3*. In 1870, a French naturalist, M Chantran, noted that in the case of *Astacus astacus* the first moult took place ten days after the hatch in early July, the second, third, fourth and fifth took place at 25 day intervals until September. There were no moults recorded from then until April, the sixth in May, a seventh in June and an eighth in July; that is eight moults in the first twelve months. In the second year two moults only took place in July and September and this was followed by an annual moult for females and twice a year for males. This pattern has since been confirmed by other people. In 1968 Jack Cukerzis produced a Table (*Table 4*) which showed the instance of moults over six years in the same species.

We have seen how the newly hatched crayfish slips out of its exoskeleton at a point between the cephalothorax and the first abdominal segment. It continues to do this at each moult but it has

28

Table 4. SEASONS AND NUMBERS OF MOULTS IN *Astacus astacus*
(*Jack Cukerzis 1968*)

Summers	Number of moults per summer	Times of moult	Age of crayfish	Number of moults per year
1st	5	1 June – July 1 Mid July – end 1 Beginning August – Mid August 1 End August – Mid September 1 Mid September – end	One year	8
2nd	5	1 May 1 June 1 July		
		1 August 1 September	Two years	4
3rd	3	1 June 1 July		
		1 End August	Three years	2
4th	2	1 June – July		
		1 August – September	Four years	Male 2 Female 1
5th	Male 2 Female 1	1 June – July ♂		
		1 August – September ♂ ♀	Five years	Male 2 Female 12
6th	Male 2 Female 1	1 June – July ♂		
		1 August – September ♂ ♀	Six years	Male 1 Female 1

29

much more work to perform in order to carry out subsequent moults. First of all it has to free all its appendages and retract them into the main part of the body. This it does by rubbing its limbs one against the other and shaking them continually. The vibrations, particularly those of the antennae can be clearly observed. This action allows some movement or play within the body and the appendages. The animal eventually splits its carapace just in front of the first abdominal segment and at this point the new soft cuticle can be seen. After frequent rests, the animal withdraws its head and, with a sudden leap forward, it will extract the whole body from the old shell. This is sometimes carried out quite quickly but normally takes a longer time, even up to 10 or more hours. Once it has started to moult the animal cannot stop and it will even continue if it is picked up. At varying stages one can feel empty parts, the easiest being the great claw. This is a period of great danger to the crayfish since it is completely helpless and can only move very slowly and so, until its new exoskeleton hardens, it is extremely vulnerable to its many enemies.

The physiological processes of the moult are controlled by very complicated hormonal interactions which are outside the scope of this book. However it is important to realize that the formation and hardening of the new exoskeleton depends ultimately on the availability of calcium. A great part but not all of this is held in the gastroliths, erroneously called *crabs' eyes*, which are chalky solids found in the stomach linings of the crayfish. They only exist in the pre-moult periods and through the moult when they are used up and finally disappear. Calcium salts are drawn from the cuticle and are transferred to the gastroliths. They can thus be considered as storehouses for calcium. At the time of moult they make up approximately 5% of the weight of the body, whereas they reduce steadily hour by hour until four or so days after the moult they only represent 0.8% ultimately disappearing to reappear in time for the next moult. As far back as 1835 it was ascertained that the composition of the gastroliths was 11.43% animal matter soluble in water, 4.3% insoluble in water, 18.6% Calcium phosphate, 63.16% Calcium carbonate and 1.41% Sodium carbonate. The gastroliths are not the only contributors of calcium for the hardening process. The hepatopancreas and the blood also play their parts and even this is not sufficient and the balance required is provided through the gills.

The hardening of the exoskeleton is therefore ultimately dependent on the amount of calcium salts in the environment.

Rate of growth

The rate of growth of the crayfish depends not only on its actual food but on its sex, species and the environment in which it lives. Males grow quicker than females. The Signal crayfish has a fairly rapid growth rate which makes it very suitable for farming. The female of this species reaches maturity in two years when it should be some 90mm in length. Waters rich in calcium are more favourable to rapid growth. It is interesting to note that *Astacus astacus* has been found to grow to eight times its own weight at hatching in the first year, after which it doubles its size every year up to five years (20–50gm). The Signal crayfish reaches the same size in two years and then grows on to even greater sizes (*Figs 8, 9* and *10*).

Fig 8   Two signal crayfish, the smaller at 0+ years, the larger at 1+. *John Hogger*

31

Fig 9 A signal crayfish. 2+ years old. *John Hogger*

Fig 10 A signal crayfish. 3+ years old. *John Hogger*

32

The regeneration of appendages

The crayfish is very susceptible to mutilation either voluntarily or involuntarily. Mutilations can be self-inflicted such as when a limb is trapped in a crevice and simply discarded so that the animal can move away, or when it is seized by an enemy and abandoned to escape. Very often the animal is mutilated at the time of mating and it is quite frequent at the time of moult when an appendage can be left in the old exoskeleton. Whatever the cause the mutilation is followed by a regeneration in which the lost part is replaced. The replacement is never as good as the original. Huxley (1881) states that limbs which are amputated during the actual period after the casting of the old exoskeleton and before the new one hardens, grow larger than the original giving rise to some monstrosities. The regeneration is closely allied to the moult. A small bud-like growth appears first of all. This grows fairly rapidly until the limb is about 25% of its original length. There follows a period of inactivity, and then it again increases in size. At the time of the moult the shell of the new limb is cast off with the rest of the exoskeleton, it acquires a new hard cuticle and eventually it is restored to its original function.

It is worth noting that the crayfish, when injured, bleeds profusely and that unless the blood coagulates and forms a skin, the animal will bleed to death.

Diseases

It is to be hoped that the devastation which can be wreaked by some of the diseases which can affect crayfish will never be experienced by the crayfish farmer. The risk is much less in the United Kingdom than in continental Europe where there is movement of stock both naturally along vast waterways and from country to country by road and rail. There are now some three areas in the South of England where there have been losses of native populations of crayfish due almost certainly to the irresponsible introduction of the Signal crayfish. This species which appears to be immune to the plague is never-the-less believed to be capable of carrying the disease. Some countries go to great lengths to prevent the spread of disease into their waters, only allowing certain specified official bodies to import crayfish other than frozen or cooked animals. Movements within countries are also the subject of legislation but this is very difficult to enforce as for example in Spain where eel fishers, seeing a new possibility for earning more, have deliberately introduced specimens from one river or paddy field to another. The importation of live crayfish into the United Kingdom is now controlled by the Wild Life and Countryside Act of 1983.

The diseases of crayfish are very complicated and little research has been carried out. They are difficult to diagnose and remedies are few and in some cases non existent. Most of the cures used generally in pisciculture are ineffective. There can be no doubt that, as in all animal husbandry and particularly in aquaculture prevention is of the utmost importance and hygiene paramount. Right at the beginning it is absolutely essential to ensure that all stocks brought into the area are disease free and the only way to be certain of this is to see that such stocks are quarantined for 40 days during which time any disease will become apparent. This quarantine should be carried out in small artificial ponds, which should be sited as far as possible from the culture area and should be capable of being drained and thoroughly disinfected before and after use. Once established, the import of further stock should no longer be necessary since the culture system should be self-supporting. All units should be closed; that is isolated one from another and all from outside. They should be secured in such a way as to prevent the entrance or exit of crayfish through the water or across land. The source of water is important and obviously the use of a spring or borehole considerably reduces the risk of infection from outside. The outflow should be so arranged that there is no pollution leaving the unit.

Food sources, implements and utensils used in food preparation must be thoroughly cleaned and disinfected. It should also be remembered that man can also be the means of introducing disease into the unit and thus facilities should be provided for the disinfection of rubber boots, *etc.* Casual visitors should be discouraged.

The three main agents of disease to which crayfish are susceptible are similar to those which affect trout and other fish namely fungi, bacteria and parasites. The diagnosis and cures are however still very much in their infancy. Amongst the fungoid diseases is the dreaded crayfish plague (*Aphanomyces astaca*) a disastrous disease which has gradually spread through Europe from America whence it was brought. It has wiped out whole populations of crayfish in Italy, France, throughout Central Europe into Scandinavia and as far as Siberia. The earliest symptoms are those of disorientation, the animal leaving the water by day and walking about as though on tip-toe. Later symptoms include white membraneous growths on the eyes and joints. Ultimately the creature will lie on its back, adopt a pedalling motion with its feet and then die. There are no known cures and various prophylactic measures involving chemicals are themselves dangerous to crayfish. The Porcelain disease (*Thelohaniosis*) is another of the killing diseases which was an additional cause of devastation in continental Europe and is caused by a microspore. The pearly lustre which appears on the muscle of the tail gives the disease its popular name. This may also be visible at the joints of the segments and appendages. Other fungoid diseases are mycosis of the eggs which causes them to turn orange coloured and eventually degenerate. A cure may be found in baths of green malachite at 10 ppm for 30 minutes at 48 hour intervals.

The possibility of identifying and curing any diseases are remote with today's state of knowledge. It is therefore essential to make and keep contact with a laboratory which can carry out tests and suggest cures. In the United Kingdom this should be found through the Fisheries Officers of the Regional Water Authorities or through the Diseases Laboratories of the Ministry of Agriculture, Fisheries and Food.

There should be a system to be followed when any abnormality is discovered in any individual crayfish. The animal should be set aside whilst a careful examination is made of every other crayfish trapped at the same time or traps should be immediately set in the area where the abnormal one was caught in an attempt to establish whether the case is isolated or part of a larger infection. Contact should be made immediately with the laboratory or veterinary service charged with the supervision of aquaculture. Careful records should be kept from the time of discovery which should include date, time and exact location within the unit of the sick animal, the symptoms and a description of the creature including movements, bearing, any markings or spots and any deformations. This record must be kept up to date with any relevant information as to the state of the water, other crayfish *etc.* Samples of the water should be taken. In consultation with the laboratory arrangements will be made for the collection and despatch to the laboratory of the infected animals, both live and dead, water samples and copies of reports and also on future arrangements for monitoring the course of the infection if this is required.

Part two
FARMING THE ANIMAL

Preliminaries

Before commencing to keep crayfish on any large scale either by an intensive method or by growing them in the wild in streams, ponds, lakes or reservoirs, there are certain preliminaries which should be considered. These fall into two categories, those which are obligatory under the law and those which are based on common courtesy.

In the former category first and foremost is the necessity to seek guidance from the Regional Water Authority both on water abstraction where this is involved, and on the disposal of effluent. These authorities have the responsibility of ensuring that no activity affects the environmental conditions by pollution *etc* in the areas under their control. Advice is freely given and should be sought on all aspects especially local water conditions and the subsoils upon which any development may be based. In many cases records will exist of water sampling which has taken place in the past as well as details of the uses to which the water has been put. If, of course, there is likely to be any large scale work it may be necessary to apply for an Impounding Licence and a Statement of Consent although this is unlikely unless it is proposed to keep crayfish in conjunction with other activities such as the creation of an artificial lake. Initial contact should be made with the Fisheries Officer at the appropriate water authority headquarters.

The second official body which may be concerned is the local council in its capacity of being the planning body whose permission has to be sought where this may be necessary. Early informal discussions with the Planning Officer are recommended since, although this officer cannot make on the spot decisions, he can recommend, and can give his views on the likelihood of what his committee may accept.

The movement of fish from one area to another, or one watercourse to another, is the responsibility of the Ministry of Agriculture, Fisheries and Food and of the Regional Water Authorities in order to prevent the spread of fish diseases and it is thought that one of these bodies may ultimately be responsible for the control of movement of crayfish. However, those who keep crayfish other than the native animal *Austropotamobius pallipes* are reminded of their obligation under the Wild Life and Countryside Act 1983 to take every practical precaution to ensure that their crayfish do not escape into natural watercourses, rivers, or streams.

The considerations which should be taken as a matter of common courtesy include such things as investigating what goes on downstream from any proposed site (and in one's own interest, upstream!); the general cleanliness of all operations including the disposal of dead bodies; the provision of inlet and outlet debris screens wherever possible and the control of supplementary feeding so that no pollution is caused. Above all it is necessary to keep proper records, especially of mortalities and diseases so that investigations into causes can be carried out efficiently and remedial steps taken. Records may prove vital to the future of what is a young industry.

The problem

The farming of all types of fish – both freshwater and marine – ornamental and for food — is carried out extensively in many parts of the world and the fish farming or aquaculture industry is growing year by year. One of the main exceptions to this is the farming of the freshwater crayfish which has received little or no attention over the years.

All the available literature presupposes a requirement for large expanses of water such as are found in the crayfish farms of Louisiana in America where many hectares are alternately flooded and drained each year, or poses problems without providing the solutions.

A handful of people in England are pioneering attempts to farm the crayfish and a number of academics, mainly biologists or zoologists, have produced papers and are carrying out research on various aspects – species, habitats, enemies, life-styles, growth rates *etc.*

The Hampshire College of Agriculture, Department of Fishery Management, headed by Dr Stephen Goddard, has provided a much needed focal point for the collection and dissemination of information on the subject, holding meetings for interested people and producing a free quarterly publication 'Crayfish Bulletin'.

In France where the crayfish is considered a great delicacy and is in much demand there have been several serious attempts to farm crayfish notably under the aegis of the Conseil Supérieur de la Pêche at sites at Vivier du Gres and Gournay-sur-Aronde, Oise. Where these have been to provide stock for populating depleted rivers and lakes they have proved successful but recent reports suggest that trials for production for human consumption have so far failed to reach a significant economic level. Efforts have been made in other

European countries, especially in Scandinavia and West Germany. All these have met with little commercial successs. The Swamp crayfish (*Procambarus clarkii*) has been introduced with some success into the rice paddy fields in Southern Spain but this cannot be called intensive growth.

Many individuals who have attempted to produce crayfish commercially have reported little success and some have been actively discouraging. A proportion of these could be considered victims of a hard sell by commercial organizations interested in the sale of juveniles for stocking purposes. In many cases this has resulted in considerable overstocking. Quite a few people, attracted by glowing publicity and the possibility of quick financial reward for little actual work, have attempted to grow crayfish in quite inadequate locations.

Perhaps the greatest amount of encouragement to pursue the problem of farming crayfish commercially comes from Alex Behrendt, the owner of Two Lakes Fishery in Hampshire, an acknowledged expert and author on the subject of aquaculture and fishery management, and above all, one of the leaders of practical crayfish rearing in this country. He wrote to me saying 'I have contacts in many countries with people interested in the crayfish, and all agree that because of the present state of our knowledge about crayfish, which is limited, farming them intensively as trout are farmed is not yet feasible. However we must not have closed minds. At Two Lakes I grow crayfish in their first summer intensively to a degree which was considered impossible a few years ago. So it may be that one day a crayfish breeder, will succeed in growing these juveniles on'.

It is Alex Behrendt's rearing ponds and brood boxes which are the basis of the early stages of crayfish farming in this book. The growing on after the first year is a theoretical study of the problems and what is believed may be the way to overcome these, problems which in the main are created by the animal itself. Field tests have yet to be carried out and the results of these will be the subject of a later book.

Considerations

The animal
Aggressiveness

The aggressiveness of the crayfish one to another, even parent to offspring, is one of the major characteristics to be considered in the context of the design of any system for rapid and large scale production. Cannibalism is very common particularly during moulting which in nature is a concerted event with most adults moulting within a short period. Juveniles, as has been seen, moult more frequently and out of phase with the adults. The cures for this, probably the principal obstacle to crayfish culture are, firstly the supply of plenty of hides, where they can have a firm and secure base; secondly, the supply of adequate food, since if they do not have to search hard for food they will tend not to eat each other; thirdly, the amount of space that each animal occupies *ie* the density of stocking; fourthly, the separation of the juveniles from adults as soon as possible, and finally the separation by size of all growing stock.

Species

This study does not propose to go into the pros and cons for raising the various types of crayfish. Suffice it to say that the closed type of production envisaged will offset many of the arguments against the introduction of the Signal crayfish (*Pacifastacus leniusculus*) and the advantages of this species – speed of growth, greater size *etc* – make it essential for purely commercial reasons. This is accepted by the most hardened of conservationists who are primarily concerned with the danger to the native species. In the long term the possibility of introducing other species such as the Australian crayfish in its various

41

forms should not be ruled out but these will require even more specialized culture involving such things as proper temperature levels and control to an even greater extent.

Depreciation of stock

It has been suggested that under normal situations, as found in nature, as many as 60 – 70% of juveniles are lost in their first year from cannibalism, escape *etc* and that out of the remainder a further 20% may be lost through predation during the growing period. This is one of the main problems to be solved before an intensive unit can be said to be successful. Cannibalism of juveniles, particularly by the parents, can be dealt with by the use of brood boxes and rearing ponds as described later. Loss by escape can be countered by the use of overhangs where necessary throughout the system, by vertical walls, or by small fences and by ensuring that close grids are provided on all outflows.

There are many predators. These include fish, notably pike, perch and eels but also other coarse and game fish, animals such as voles, otters, mink *etc.*, and birds, especially the heron, moorhen and coot. Some larva, including that of the dragonfly, will eat juveniles voraciously.

It is believed that a completely closed system will deal very efficiently with all of these problems, since parts, if not all, of the system can be effectively netted.

Water quality

Control of water should be comparatively easy. Points to note are pH values of 7 upwards (7 – 9 preferable), hardness of between 100 and 350 ppm $CaCO_3$, a minimum level of 6 ppm of dissolved oxygen. Supplementary aeration of parts of the system may be desirable and even essential on occasions. The temperature of the water does not appear to be critical in that crayfish will survive in extreme cold in winter provided that the water does not freeze right through. This can be assumed with moving water if the depth is greater than two feet in the United Kingdom. There must be a minimum of 15°C for at least three months in the summer, and early hatching can be achieved by providing warm water from January onwards of about 10 – 13°C.

Feeding
Serious consideration and thorough research will be necesssary to find out the amount of supplementary feeding that is required since the system will be artificial throughout and the stock will not be put into open lakes where crayfish can normally fend for themselves as far as food is concerned. This is important not only for survival but also for rapid growth to marketable size. Large quantities of weed, especially blanket weed, will be provided in the growing units and there will be a supply of snails and other small crustacea *etc.*

Disease
With a closed system as envisaged, control of the environment will be easier and thus the elimination of disease. The points at risk are the source of the water and the outflow. It would be preferable to use water from a borehole and possibly to recycle it through the unit. Consideration should be given as to what precautions and measures can be taken to clean any water that is obtained from an outside source such as a stream although risks should be fairly minimal. This may become an increasing problem as more and more people take to setting up crayfish establishments. As regards the outflow a settling tank and possibly a sand/gravel filter should suffice to ensure that the discharge from the system is reasonably clean. The fact that the greater part and possibly all the food to be provided may be live, either animal or vegetable, will of course help in the effluent problem.

It is recommended that a perimeter fence should contain the whole unit not only to provide security from poaching but also to keep out unwanted sightseers, thus reducing the amount of supervisory work required in sterilizing boots *etc.* All equipment, including the footwear of operators, should be sterilized and casual visitors should be discouraged.

Mating

This is a most problematical area and one where the operator can influence matters to a limited extent. In the wild crayfish mating takes place in late October. The females then retire and the gestation periods qv take place. If they are disturbed the chances are that the eggs will not be stuck on to the pleopods and are found free in the water. Egg laying occurs in the first two weeks in November.

It is suggested that if they are left to accept a cold shock (0 – 4°C) for a week or two and then brought inside the eggs will mature and the young will arrive earlier in the season – say March as against June or July. This might enhance the growth rates. Others could be left outside to provide a crop of juveniles later in the year. This might provide a more continuous production pattern. It would appear therefore that at least one of the rearing ponds should be under cover where water temperature can be efficiently and economically controlled. The other consideration in this connection is one already referred to, that is the advisability of ensuring that stock crayfish used for breeding purposes are all of a similar size and are kept in the proportion of one male to every three or four females.

Harvesting

Although traps will be retained for monitoring purposes, the closed system will obviate the need for trapping since it will be possible to drain and collect harvestable crayfish. The harvestable size will be 10 centimetres minimum. The measurement is taken from the tip of the rostrum to the end of the tail. The possibility of grading must be considered, since the market price will vary considerably according to the size of the crayfish.

Components of the system

The brood box

The object of the brood box is to prevent the mother being able to eat her newly born baby crayfish. It consists of a simple four-sided wooden box, approximately 40 cm square with a divider down the centre. The top and bottom are covered with rigid netting of approximately 13 mm mesh. The top netting must have some form of opening which can be firmly secured.

The box, illustrated at *Fig 11* contains a piece of plastic pipe about 5 cm in diameter and about 17 cm in length, and also a piece of flat tile or plate to each side.

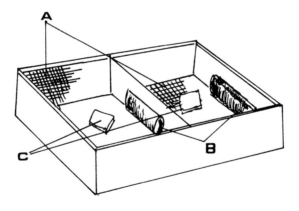

Fig 11　The brood box. A – mesh to allow passage of juveniles. B – plastic tubes to provide hide for mother. C – flat plate or tile for food.

As close to the hatching date as possible a female with eggs is placed into each side of the box. The pipe provides a suitable hide and the tile enables food to be placed in the box without it falling through the bottom mesh.

The brood boxes (or rather a series of them) are placed in the rearing ponds, raised on bricks to allow a space under each so that the juveniles can get out through the bottom mesh.

Eventually the juveniles will hatch, remain with the mother until they are free swimming and then they will fall through the bottom mesh into the rearing pond where the mother cannot get at them. At this stage the mother is removed and returned to the stock pond for future breeding (or harvesting). The mother must not be left with the young under any system for more than 20 days. When the young have left the brood boxes and the females have been returned to the stock pond, the boxes are also placed in the stock pond to ensure that any juveniles that have been missed can get out and fend for themselves.

The rearing pond

The rearing pond, illustrated at *Figs 12, 13* and *14*, is a long concrete pond with a loose gravel base. The length depends on the number of juveniles it is proposed to rear in it, and it is considered better to have several ponds rather than one very long one since this makes for ease of handling. It is approximately one to two metres wide, and about

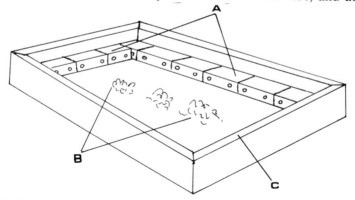

Fig 12 The rearing pond. A – air bricks placed round the edge. B – extra hides. C – overhang to prevent escape.

46

30 cm deep. On the edge all round there must be an overhang or else a small fence to prevent the juveniles from escaping. Alternately the whole pond can be netted using one of the rigid nylon nets which are commercially available.

Fig 13   One of the rearing ponds at Two Lakes Fishery, Romsey, Hants. The air bricks and the additional hides can be clearly seen. *John Hogger*

Fig 14 A rearing pond at Two Lakes Fishery with the nets in position to prevent predators. *Alex Behrendt*

All round the edge of the pond in the water air bricks are placed to provide hides for the juveniles and extra hides are scattered all over the bottom at random. These can be of chopped up hose, plastic piping, broken air pipes or even strips of corrugated plastic sheeting – anything which will provide cover. It has been suggested that domestic pot or pan scrubbers such as are made of loosely woven plastic in the shape of a ball are excellent for providing hides for the juveniles.

Whatever material is used the hides should never be provided on a scale of less than three or four per potential juvenile. It is better if they can choose amongst a surplus of hides and also if they cannot see each other the whole time.

Provision must be made for covering the whole pond with netting to keep out predators and, should the juveniles be kept in the pond for any length of time, a double layer of polythene sheeting may be necessary to keep out the frost.

The water in these ponds must be flowing but this can be at quite a slow rate.

The juveniles are not regularly fed but blanket weed and natural pond life (net swept) are placed in the pond at the beginning of each season after the ponds have been emptied and cleaned out ready for use. Occasionally trout pellets can be fed but these must be removed if uneaten to prevent any pollution of the water.

The water level can be kept just above the top of the air bricks *ie* 15 cm or so, and this level can be increased as the juveniles grow. Juveniles can be periodically thinned, holding some in the rearing ponds at a lower density since otherwise these ponds will be empty for some six months of the year.

If the rearing ponds are not themselves netted as suggested above, then provision must be made for overall netting to keep off predators.

Referring to the previous remarks on the growth rate of crayfish it is possible that at least some of the rearing ponds should be capable of being heated in some way – this could be in the form of glasshouse type covers, polythene tunnels and/or the possibility of some form of solar heating.

Summerling unit
When the crayfish reach about 3 – 4 cm in length after approxi-

mately four months they could be removed from the rearing ponds and kept in a further growing stage – a summerling unit. This is the stage when they would normally be placed in open lakes to fend for themselves. For the purposes of this study it is proposed that they will be kept in the rearing ponds a little longer – until they are 6 months old. They will then be transferred into the main growing unit, where they will be segregated by age and size until they are harvestable.

Growing unit

So far we have been using systems which have already been tried and proved to be successful. The growing unit is the crux of this suggested intensive method of production.

It is usual to grow on the summerlings by putting them into natural habitats such as lakes and reservoirs where they are left more or less to their own devices to fend for themselves and grow. They are then trapped and sold when they are of a marketable size. This would appear to be the real position of risk in the whole operation of crayfish production; they are very susceptible to all the dangers – predation, pollution, poaching *etc* not to mention each other. Furthermore the task of trapping is haphazard and time consuming and given numbers cannot be guaranteed.

Although crayfish do venture out into deep waters, it is believed that they are primarily a shallow water creature and a great deal of space is wasted in growing them in open lakes. Indeed it is possible that only 20% of any open stretch of water is utilized by the crayfish. This may be because the majority of hides are in shallow water and certainly they tend to move into shallower and consequently warmer water in the Spring. It is on this premise that the whole of this present theory is based. The principles are shown in *Figs 15* and *16*. If the crayfish can be asssumed to live perfectly well in shallow water, that is to say the bank areas of the lakes, then this can be created artificially with the tremendous advantage of space-saving, control from pollution, ease of harvesting – in fact complete control over all aspects of the environment.

The following points should be noted:

(1) The waterway is approximately 2 metres wide.

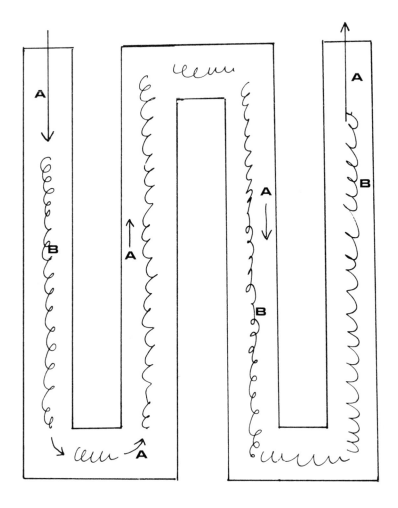

Fig 15   The theory behind the growing unit. A – direction of flow. B – hides

51

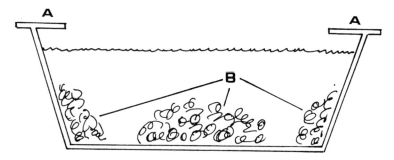

Fig 16 A cross section of the growing unit showing A – the overhang. B – hides along each edge and in the centre of the channel

(2) The depth of water must be above 60 cm to obviate total freezing in winter.

(3) A good bed of gravel on the bottom of the channel will allow a rich microfauna to become established which will provide an additional food source.

(4) The rate of flow need only be minimal – but there must be a flow.

(5) Both banks and the centre of the channel are provided with hides made of rubble *etc*. The use of limestone rocks is suggested since this will help with the hardness of the water.

(6) Provision should be made to drain each section separately and completely – this is essential for harvesting, rendering trapping unnecessary. It will also enable cleaning to be carried out efficiently.

(7) There must be a lip or overhang to prevent escape.

(8) Grids can be inserted at intervals in the unit to allow separation and grading by size and/or age thus allowing complete control of stock. Furthermore it should be constructed to allow for short circuits of sections to allow for the isolation of any problem areas as may be necessary from time to time.

(9) Provision can be made for the complete control of water entering the system.

(10) The whole area, being compact, can be netted to keep out predators.

The stock pond

The growing unit will render large expanses of water unnecessary. In

52

addition it will be necessary to make provision for the preservation of the crayfish which will be used for breeding purposes and thus provide input to the system. This unit which we will call the stock pond could be a natural piece of water but preferably and in order to keep the whole system closed it will be a man made pond or pool. It can be quite small since it will only hold the brood stock.

This breeding stock will consist of males and females of breeding age in the proportion of one male to every three or four females. The sexes will be of similar size for the reasons set out in the earlier chapter on the mating of crayfish. These crayfish will be marked and recorded so that if it proves possible the strain can be improved. From each year's crop several of the best will be kept back for breeding purposes.

Since the crayfish is very prolific and with the reduction in the depreciation of stock which this system puts forward numbers can be kept fairly small. This in turn governs the size of the stock pond required since the rate of occupancy need not be more than 120 per acre.

It will be preferable if the stock pond can be drained in common with all the rest of the system. This will render trapping unnecessary.

In natural waters feeding should not be required, but in a completely artificial environment, supplementary feeding may be necessary. When the crayfish are taken out for removal to the brood boxes the opportunity should be taken to clean out the pond, replant it with aquatic weed and water life, especially snails, ready for their return. The males will have to be kept elsewhere whilst this annual spring clean is carried out.

The production unit

A schematic lay-out, not to scale, for the basis of a production unit is shown at *Fig 17*.

The area of the stockpond is 90 square metres. This will be adequate since it will be able to support up to 90 crayfish of breeding size – more than will be required for the unit. The combined area of the two rearing ponds is 24 square metres. Once in operation it may be necessary to increase the number of rearing ponds to provide sufficient stocks of juveniles for the growing units. The total area of these

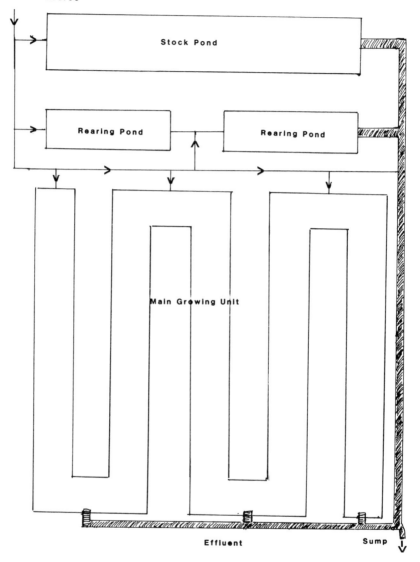

Fig 17   The layout of a production unit

54

is 1,036 square metres. This area should support a total of at least 2,500 growing crayfish and if the intensive system proves successful many more. (It has even been suggested that up to 100 crayfish weighing up to 25 grammes each could be accommodated in hides of one cubic metre total volume!)

Allowing for five metres of space between the stock pond, the rearing ponds and the growing units and with a similar space all round, this will take up about one acre in space. This is very generous in unproductive space and by reducing the areas allocated between units the whole system could be accommodated without undue over-crowding.

It is suggested that one production unit should consist of a stock pond, six rearing ponds and three growing units. Further production units should be completely self-supporting, that is, each with their own components to ensure adequate control of the units, a barrier against disease and, in the event of any unforeseen disaster, an isolation of the unit.

Since water flows throughout need not be very great all the pipework can be of fairly small bore and can be ordinary commercial PVC manufacture.

Food for crayfish

There are four main types of food for crayfish which can be classified as live food, dead food, vegetable matter and artificial food. Contrary to popular belief the crayfish, even in the wild, does not go out of its way to seek out dead and rotting animal matter or vegetation to eat, preferring fresh food and only consuming such stale substances when other food is in short supply or not available.

There have been many experiments conducted, especially in America, in an attempt to find the best types of food. These include experiments involving listening to the animals crunching various foods with hydrophones to see which they like best. Out of all these experiments several basic guide lines have been established.

(1) Crayfish can be very individual in their diets and therefore in large numbers there is some advantage in a mixed diet so that all can be happy.

(2) Crayfish consume amounts in direct proportion to their body weights depending on their age, sex and seasonal changes.

(3) Food must be given in the right quantities to ensure that there is sufficient on the one hand and that it is all consumed on the other to avoid pollution of the water.

(4) During periods of intense activity, particularly during the mating season, food consumption rises steeply.

(5) Crayfish eat primarily at night, therefore feeding should take place in the late afternoon. They should, however, never be without food entirely so that they look hungrily at their neighbours!

(6) As crayfish grow they will consume more vegetable matter and less animal.

Live food
The principal live food for crayfish are snails of all types but mainly the wandering snail (*Limnea peregra*), the great pond snail (*Limnea stagnalis*) and the ramshorn snail (*Planorbis planorbis*). It is very fascinating to watch a crayfish ease out a snail very gently from its shell, but usually they crush the shell between their large claws and eat the animal within. They have been observed to eat the whole animal, shell and all, and this may possibly provide a further source of calcium. Water snails can be very easily cultivated in a separate small pool and they are very prolific, thus providing a good supply of food for the growing crayfish. They can also be a good side crop for supplying local pet shops and garden centres who then sell them on to ornamental pond and aquaria owners.

In addition to those referred to above there are many other varieties of snail to be found in local ponds and streams which can be collected and cultivated.

One interesting note about snails. Their droppings, unlike those of fish, do not decompose and therefore can be the cause of pollution in closed areas.

Another live food for crayfish is the earthworm which they will eat avidly. These can also be raised in great quantities in boxes or more permanent constructions. Great care must be taken to ensure that all earthworms fed are in fact consumed since any left will drown and will then cause pollution. Other types of worm, especially the tubifex which can exist in the habitat, are thus much safer to use.

Contrary to what may seem to be a natural food, other crustaceans found in freshwater such as the freshwater shrimp (*Gammarus pulex*) are not suitable since the crayfish will not be able to catch them. The only exception to this being very small creatures, daphnia, rotifers *etc* which can be net-swept from other waters and put into rearing pools for juveniles.

Dead food
When establishing a breeding centre for crayfish, investigations should be carried out into other local activities to see if there are any likely sources of food. Great successes have been achieved by the utilization of waste products from fish canneries *etc*. Noteworthy as an example is where there was a plant bottling mussels, the refuse being

in the form of shells with scraps of the animal still attached. Such foods can be used for crayfish with great success and little expense. At the risk of labouring the point it is important to remember that these dead foods are potential pollutants if they are not wholly consumed. They can be fed on lines or nets so that unconsumed portions can be easily removed.

Artificial foods

Unlike all other fields of aquaculture there are so far no artificial foods in pellet or similar form especially available for crayfish mainly because of the lack of demand. There is some work going on in this field especially in connection with feeds for shrimp production. When the demand comes no doubt the effort will be increased and solutions found.

Several formulae have been produced (Meyers *et al* 1972) on an experimental basis, but so far there has been no commercial production. Crayfish will eat trout pellets where nothing else is available, indicating that it may be possible to use a dry feed pellet as the sole type of food if one could be designed with the dietary requirements of the animal.

Vegetable foods

Crayfish will eat and grow on almost every kind of vegetable including the various beets and it may be possible to find a local source in the same way as suggested above from local processing plants.

They will flourish on practically every type of aquatic plant; indeed they are an essential part of their diet. Their favourites are undoubtedly the various varieties of *Elodea* and ordinary blanket weed. These can also be cultivated in separate pools or old tanks and can be transferred to the main growing units after each cleaning and from time to time as required. There they will grow (until eaten!) and will have no pollutant effect. They will also provide a useful side crop for sale to pet shops, garden centres and the like.

How much food?

The amount of food that a growing crayfish consumes is well established. Juveniles eat between 1 and 4% of their total body weight each day, whereas adults consume less at between 0.3 and 1% per day. This latter figure can be sub-divided further in that, during the mating season, females will eat between 0.3 and 0.4% with males consuming 1% whereas, outside the mating period they consume 0.3% and 0.6% respectively.

It is therefore possible to calculate the amount of food to be provided each day according to the numbers of individuals, their sexes and the growth of the stock in various parts of the system. This is a delicate balance to be maintained between providing sufficient food to ensure a continued and steady growth on the one hand and to make sure that all food that is provided is consumed each day to avoid the risk of pollution on the other. When this balance by weight has been determined then the decision has to be made as to how this is to be broken down between the various types of food that are available, including an increase in vegetable material as the stock grows older. Careful records at various stages of growth should be kept which should show the date, the weight of the crayfish, the weight of each type of food and the frequency of supply. These records should be kept on a weekly basis, necessitating the capture and weighing of several individuals from each part of the system to establish an average weight.

Feeding must be carried out systematically and in strict accordance with a programme and properly recorded. The effects of this programme must be carefully noted and amended as required until the correct balances referred to can be achieved.

When the programme has been finally proved by experience it should be adhered to. Monitoring should take place to ensure that all is well and to enable some modifications to be made if there are changes in basic conditions, such as may occur during a long spell of hot weather causing a rise in water temperature which, if sustained, may cause an increase in the rate of growth.

Growth target
This should be in accordance with *Fig 18*. It will be noted that the minimum marketable size of 10 cm should therefore be reached in three years. *Fig 19* shows the relationship weight/length. For those with a mathematical bent there is a formula for establishing a ratio of weight to length based on the following equation:

Males $\text{Log W} = -5.0773 + 3.5920\,(\text{Log L})$
Females $\text{Log W} = -4.6003 + 3.0726\,(\text{Log L})$

Where W is the bodyweight in grammes and L is the length in millimetres.

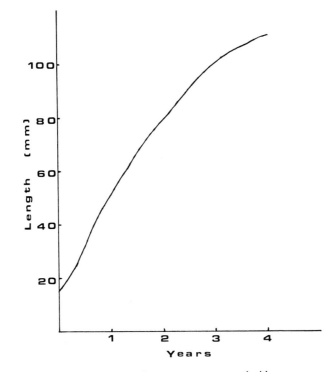

Fig 18 Curve showing the length of the carapace compared with age.
Signal crayfish

Conclusions

From the foregoing it will be seen that an ideal system would be one where the juveniles are transferred from the rearing ponds into growing units which have already been stocked with a known weight of growing aquatic plants and water snails of various types. That this live food should be replaced in toto if possible, or supplementary foodstuffs used according to a programme or regimen that has been worked out in advance, tested by experience and found to produce a steady growth rate culminating in marketable crayfish at the end of the third year, or sooner if this can be achieved.

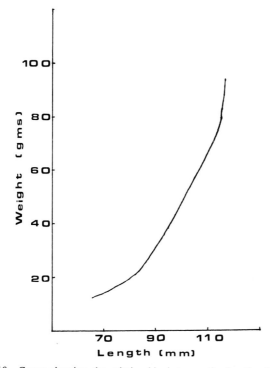

Fig 19 Curve showing the relationship between the length of the carapace and the body weight. Signal crayfish

Keeping crayfish in open waters

The scheme which is outlined in Part Two of this book is an attempt to show the possibilities of rearing the crayfish intensively for the market. It is, of course, possible to rear crayfish in almost any stretch of water be it lake, pond, gravel pit, reservoir or stream, provided it has suitable conditions which broadly are set out earlier. *Figs 20* and *21* show the use of a trout stew pond.

The best way to stock such waters is by the introduction of juveniles in the late spring. These can be bought commercially from a variety of sources and it is now unnecessary to import them. Another

Fig 20   Introducing a trial stocking into a stew pond on a trout farm. *John Hogger*

Fig 21 Collecting crayfish from the stew pond in Fig 20, drained down. *John Hogger*

method is to introduce female crayfish which are already carrying eggs.

The losses which can be expected in either case are very large but more so in the latter since the young are at risk from their mothers. For example if a start was to be made with say 1,000 juveniles it would be very fortuitous if there were 600 at the end of the first year; if a similar loss occurred during the second year, this would leave only 300 to commence the breeding cycle, some male and some female. Admittedly the females, for the sake of argument 50%, that is 150, could produce some 15 – 20,000 young. However the mortality rate in these would be even higher, probably leaving only one in a hundred to grow through to two-year-olds. These losses would be through cannibalization, losses during moults and losses through predators of all kinds. Wherever possible steps should be taken to counteract these losses – for instance by the provision of extra hides along the edges of the water, the planting of additional weed as cover, the removal of predatory fish which may eat the small juveniles, constant attention to the entry of eels, animals such as the mink and birds such as the heron.

It can be possible to combine some of the features envisaged in an intensive unit with open lakes and indeed this is done in several places already. There is no reason why the brood box and the rearing pond should not be used to get young crayfish through the early and dangerous periods and then let them grow on in open water, by and large fending for themselves. Some supplementary feeding may be advisable since this may help to provide more rapid growth.

Crayfish kept in open waters will have to be trapped to harvest them. Care should be taken to ensure that no animal is taken out under 10 cm in length and, for the first few years, only males should be taken if a population is to be established. This is because one male can mate with several females and is therefore more expendable! It is wise to mark a certain number of males and to return these to the water in order to ensure that some males are left. If re-caught later they can then be returned again.

Traps can be bought ready made, some plastic and collapsible for easy transport and storage. They can be easily made following the principles of the lobster pot on a smaller scale and using small mesh chicken wire or plastic netting (*Fig 22*).

Fig 22 Crayfish traps. The brick shows the approximate dimensions. *John Hogger*

One of the problems of keeping crayfish in open waters is in the consistency of numbers caught and the ability of being able to guarantee to the purchasers that a given number can be supplied on a regular basis. It is therefore necessary to be careful to keep any regular customers within the limits that can assuredly be supplied.

If all the precautions contained in this book are carried out the rearing of crayfish 'in the wild' can be rewarding and fun. It will not be a medium for the fortune-hunter as has sometimes been implied. It will, however, provide a useful addition at little expense to the farmer's table and may produce a limited surplus for sale to local restaurants. Some may do better and produce a steady supply but growing the crayfish in this manner in this country will never succeed in the same way as the farming of trout has in recent years.

Cooking the crayfish

In England recipes for cooking crayfish are few and far between since they are a comparatively unknown delicacy. It is to France that one has to go to seek out the finer points since in that country the crayfish is considered a great subject for the culinary art. Madam Prunier (1967) lists some 16 methods of cooking the creature and it is to her that the real bon viveur should turn. Lesley Morrisey (1978) has produced an interesting book of West Australian recipes, mainly concerned with the large Marron crayfish but some of which could be tried.

The recipes which appear below will suffice for most people. The crayfish should be killed by immersing it in fast boiling water when it will die instantly. To prepare it for cooking rinse it well and remove the intestinal tract by pulling out the middle tail fin (*telson*) when it should come out whole. Crayfish are a greenish brown colour but turn bright red when cooked.

Boiled crayfish
This recipe and the slightly more complicated one which follows can be used to cook and serve the crayfish either hot or cold. If the latter, the crayfish should be allowed to cool in the liquids in which it is cooked. The crayfish should be boiled for 2 minutes in a well-flavoured fish stock and then simmered for about 10 minutes. Variants of this are the methods used extensively throughout England for cooking the native crayfish. Instead of a simple fish stock, what is known as a court bouillon may be used.

Ingredients

4 – 6 whole crayfish	A bouquet garni
2 onions	6 peppercorns
1 carrot	The juice of half a lemon
20 gm butter	3 cups water
1 cup dry white wine	Teaspoonful salt

Method

Slice the carrot and the onions. Fry gently in the butter until golden brown. Add the wine and the water slowly. Add all the remainder and simmer gently for 15 – 20 minutes. Put in the crayfish. Boil for 2 minutes and then simmer for 10 minutes. If to be served cold allow to cool in this liquid.

Écrevisses à la Bordelaise

This is the traditional French recipe for cooking crayfish which are served whole and unshelled. The dish can be prepared the day before and re-heated gently. To serve four people.

Ingredients

12 – 16 whole crayfish	3 cups dry white wine
4 – 6 carrots	2 tablespoonfuls tomato paste
4 onions	8 tablespoonfuls cream
300 gm butter	Salt, black pepper and
8 sprigs parsley	cayenne pepper to taste
2 bay leaves	

Method

Chop the vegetables into thin strips (julienne). Melt the butter in a larger pan and simmer the vegetables and herbs until the former are soft. Add the crayfish and cook over high heat, stirring continually until the shells are red. Add the wine and tomato paste and boil gently for a further 15 minutes. Add the cream. Take off the stove and season to taste.

Fried crayfish

Crayfish may be fried or grilled using butter as the medium. The following recipe will improve them immensely.

Ingredients
4 – 6 whole crayfish
1 onion   1 tablespoonful tomato paste or puree
1 small carrot   2 wineglassfuls dry white wine
2 sprigs parsley   60 gm butter
1 bay leaf   A pinch of powdered thyme
A bouquet garni   Seasoning to taste

Method
Cook all the ingredients together in half the wine and butter until the vegetables are quite cooked. Add the rest of the butter and then fry the crayfish until the shells turn red. Add another glass of wine, the tomato paste or puree, a bouquet garni and seasoning to taste. Cover the pan and cook for a further 15 minutes. Arrange the crayfish on the plates, thicken the sauce and pour it over the crayfish. Garnish with parsley.

Smoking
For those interested and who have the necessary equipment it is possible to smoke crayfish. This can be carried out in a small home smoker. For those with more sophisticated equipment the following method should be adopted.

Method
In either case the crayfish should be peeled and put into a basic fish brine. The time will vary, according to size, from 15 – 45 minutes. Remove from the brine, rinse lightly in fresh water and put on racks or perforated aluminium foil to dry.

Begin smoking at about 30°C. After 15 minutes increase the temperature gradually to 55°C.

After 60 – 90 minutes the crayfish should have taken on a rich golden colour. Taste one of medium size to see if it is cooked. Remove them or leave a little longer as may be required and note the size/temperature for future use. They may be served hot or cold. After smoking the crayfish may be oiled which gives them a richer flavour. Put them into a container with a screw-top lid. A pickle or coffee jar is ideal. Add some olive or cooking oil and screw down the lid. Lay the jar on its side and keep rotating it from time to time until all the crayfish are coated with oil. This oil will be absorbed into the flesh.

Rotate the jar again to redistribute the oil and continue turning at intervals until the crayfish will absorb no more. This may take an hour of intermittent attention.

Crayfish which are smoked and oiled in this fashion will keep in a refrigerator for several days or may be bottled or canned.

References

Animal Biology. Grove, A. J. and Newell, G. E. (1974) University Tutorial Press, Cambridge.

Alginates as Binders for Crustacean Diets. Meyers, S. P., Butler, D. and Hastings, W. (1972) and Meyers, S. P. and Zein, Eldin Z. (1972) Proceedings of the 3rd Annual Workshop. World Mariculture Society. 351–364.

Crayfish Bulletin. Hampshire College of Agriculture, Dept of Fishery Management.

Étude descriptive des principes étapes de la morphogénése sexuelle chez un crustace decapode. Payen, C. (1973) C.N.R.S. Ann. Embryol. et Morphogénése VI. 2:181–185.

L'Écrevisse et son Élevage. Arrignon J. (1981) Gauthiers-Villars, Paris.

Madame Prunier's Fish Cookery Book. Ed. Ambrose Heath (1967) Hutchinson.

Observations sur l'Histoire naturelle des Ecrevisses. Chantron, M. (1870) C.R.Ac. Sciences T 71: 43.

Proceedings of the First International Symposium on Freshwater Crayfish. Sweden 1974.

Proceedings of the Second International Symposium on Freshwater Crayfish. USA 1976.

Proceedings of the Third International Symposium on Freshwater Crayfish. Finland 1978.

Proceedings of the Fourth International Symposium on Freshwater Crayfish. France 1980.

Proceedings of the Fifth International Symposium on Freshwater Crayfish. USA 1983.

Proceedings of the Sixth International Symposium on Freshwater Crayfish. Sweden 1984.

Proceedings of the 2nd Annual Texas Crayfish Production and Marketing Workshop. Jay V. Huner.

Relationship between body weight and the age of *Astacus astacus*. Cukerzis, J. (1968) Limnology. 3 (2). 124–129.

The Crayfish. An Introduction to the Study of Zoology. Huxley, T. H. (1880) Paternoster Press. *Reprinted* The MIT Press (1973).

Western Australian Crayfish Cookery. Morrissey, Mrs L. (1978) Claire Dane, West Australia.

Other books published by
Fishing News Books Ltd

Free catalogue available on request

Advances in aquaculture
Advances in fish science and technology
Aquaculture practices in Taiwan
Aquaculture training manual
Atlantic salmon: its future
Better angling with simple science
British freshwater fishes
Business management in fisheries and aquaculture
Commercial fishing methods
Control of fish quality
Culture of bivalve molluscs
Echosounding and sonar for fishing
The edible crab and its fishery in British waters
Eel capture, culture, processing and marketing
Eel culture
Engineering, economics and fisheries management
European inland water fish: a multilingual catalogue
FAO catalogue of fishing gear designs
FAO catalogue of small scale fishing gear
FAO investigates ferro-cement fishing craft
Fibre ropes for fishing gear
Fish and shellfish farming in coastal waters
Fish catching methods of the world
Fisheries of Australia
Fisheries oceanography and ecology
Fisheries sonar
Fishermen's handbook

Fishery development experiences
Fishing boats and their equipment
Fishing boats of the world 1
Fishing boats of the world 2
Fishing boats of the world 3
The fishing cadet's handbook
Fishing ports and markets
Fishing with electricity
Fishing with light
Freezing and irradiation of fish
Freshwater fisheries management
Glossary of UK fishing gear terms
Handbook of trout and salmon diseases
Handy medical guide for seafarers
How to make and set nets
Introduction to fishery by-products
The lemon sole
A living from lobsters
Making and managing a trout lake
Marine fisheries ecosystem
Marine pollution and sea life
Marketing in fisheries and aquaculture
Mending of fishing nets
Modern deep sea trawling gear
Modern fishing gear of the world 1
Modern fishing gear of the world 2
Modern fishng gear of the world 3
More Scottish fishing craft and their work
Multilingual dictionary of fish and fish products
Navigation primer for fishermen
Netting materials for fishing gear
Pair trawling and pair seining
Pelagic and semi-pelagic trawling gear